SNOW WHITE IN NEW YORK

FIONA FRENCH

OXFORD UNIVERSITY PRESS
OXFORD MELBOURNE TORONTO

Oxford University Press, Walton Street, Oxford OX2 6DP

Oxford New York Toronto
Delhi Bombay Calcutta Madras Karachi
Petaling Jaya Singapore Hong Kong Tokyo
Nairobi Dar es Salaam Cape Town
Melbourne Auckland

and associated companies in
Berlin Ibadan

Oxford is a trade mark of Oxford University Press

© Fiona French 1986
First published 1986
Reprinted 1988, 1989
First published in paperback 1989

British Library Cataloguing in Publication Data

French, Fiona
Snow White in New York
I. Title
823'.914[J]   PZ7

ISBN 0–19–279808–1 (hardback)
ISBN 0–19–272210–7 (paperback)

dedicated to Molly and Gordon,

Brian and Jennifer

Set by Oxford Publishing Services, Oxford
Printed in Hong Kong

Once upon a time in New York there was a poor little rich girl called Snow White. Her mother was dead and for a while she lived happily with her father. But one day he married again . . .

All the papers said that Snow White's stepmother was the classiest dame in New York. But no one knew that she was the Queen of the Underworld. She liked to see herself in the New York Mirror.

But one day she read something
that made her very jealous

'Snow White
the Belle of New York City.'

And she plotted to get
rid of her stepdaughter.

'Take her down town and shoot her,'
she said to one of her bodyguards.

The man took Snow White deep
into the dark streets, but he
could not do it.

He left her there,
lost and alone.

Snow White wandered the streets
all night, tired and hungry.
In the early morning she heard music
coming from an open door. She went inside.

The seven jazz-men were sorry for her.
'Stay here if you like,' they said,
'but you'll have to work.'

'What can I do?' she asked.
'Can you sing?' said one of them.

The very first night Snow White sang
there was a newspaper reporter in the club.
He knew at once that she would be a star.

Next day Snow White
was on the front page of the
New York Mirror.
The stepmother was mad with rage.
'This time I shall get rid of her
myself,' she said.

And so she decided to
hold a grand party in honour
of Snow White's success . . .
but . . .

. . . secretly she dropped a poisoned cherry in
a cocktail and handed it to Snow White
with a smile.

Crowds of people stood
in the rain and watched
Snow White's coffin pass by.

The seven jazz-men,
their hearts broken,
carried the coffin
unsteadily up the
church steps.

Suddenly one
of them stumbled,
and, to everyone's
amazement,

Snow White
opened her eyes

The first person she saw was the reporter. He smiled at her and she smiled back. The poisoned cherry that had been stuck in her throat was gone.

She was alive.

Snow White and the reporter fell in love.
They had a big society wedding,
and the next day cruised off on
a glorious honeymoon together.